STEM in the MILITARY

MISSILES in Warfare

By Tanner Billings

Cavendish Square

Published in 2023 by Cavendish Square Publishing, LLC
2544 Clinton Street, Buffalo, NY 14224

Website: cavendishsq.com

Library of Congress Cataloging-in-Publication Data

Names: Billings, Tanner, author.
Title: Missiles in warfare / Tanner Billings.
Description: Buffalo, New York : Cavendish Square Publishing, [2023] |
Series: STEM in the military | Includes index.
Identifiers: LCCN 2022027029 | ISBN 9781502665522 (library binding) | ISBN 9781502665515 (paperback) | ISBN 9781502665539 (ebook)
Subjects: LCSH: Guided missiles--Juvenile literature. | Ballistic missiles--Juvenile literature.
Classification: LCC UG1310 .B55 2023 | DDC 358.1/7182--dc23/eng/20220608
LC record available at https://lccn.loc.gov/2022027029

Editor: Kate Mikoley
Copyeditor: Abby Young
Designer: Andrea Davison-Bartolotta

The photographs in this book are used by permission and through the courtesy of: Cover yuri4u80/Shutterstock.com; p. 5 bibiphoto/Shutterstock.com; p. 6 U.S. Space Force photo by Anthony Mendez, 30th SCP Mobile Optics/DVIDS; pp. 9, 35 Sergey Kohl/Shutterstock.com; p. 11 Peter Etchells/Shutterstock.com; p. 14 wikimedia commons/File: BGM-109G_Launch.jpg; p. 16 Prachaya Roekdeethaweesab/Shutterstock.com; p. 19 sibsky2016/Shutterstock.com; p. 20 courtesy of Archives.gov; p. 22 U.S. Navy photo by Lt.j.g. Samuel Hardgrove; p. 24 photo courtesy of Tech. Sgt. Andy Dunaway/U.S. Army; p. 27 U.S. Navy photo by Maj. Nicholas Chopp; p. 28 U.S. Navy photo by Mass Communication Specialist 3rd Class Joe Bishop/Released/DVIDS; p. 30 Andrew Harker/Shutterstock.com; p. 33 Joseph Sohm/Shutterstock.com; p. 34 Photo by Staff Sgt. Sharon Singer/DVIDS; p. 36 U.S. Air Force photo/Airman Collin Schmidt; p. 38 U.S. Air Force photo; pp. 40–41 U.S. Air Force photo by Airman 1st Class Danny Monahan.

CPSIA compliance information: Batch #CWCSQ23: For further information, contact Cavendish Square Publishing LLC at 1-877-980-4450.

Printed in the United States of America

CONTENTS

INTRODUCTION

When you think about weapons used by the military, you might think about guns, fighter jets, bombs, and missiles. But do you know what a missile really is? There are actually several different kinds of missiles, but in general, the term refers to any weapon that is **propelled** by a rocket and made to send an explosive warhead to a certain location. Missiles are known for traveling at high speeds and having great accuracy. Most missiles are controlled in some way, and for that reason, they are commonly referred to as guided missiles.

Missiles come in a wide range of sizes and distances that they can effectively travel. There are underwater missiles, called torpedoes, and low-flying guided missiles known as cruise missiles. A **ballistic missile** is a missile that travels in a high arc and is guided and rocket-powered for only the ascent of its journey. Once a ballistic missile begins its descent, it is no longer powered or guided.

In this book, you'll learn more about the different types of missiles, their history, and how they're used by the military. You'll also learn about different jobs associated with missiles in

A military missile that is not guided is typically referred to as a rocket.

warfare and how a good working knowledge of STEM—science, technology, engineering, and math—is important if you ever want to have one of these jobs. Read on to find out more about missiles and their role in warfare.

Today's military missiles are more advanced than ever before.

CHAPTER 1

EARLY MISSILES

Like most military technology, missiles have evolved and become more advanced as the technology involved in building and using them has improved. Today's military missiles are the most advanced missiles that have ever existed, and it's likely even more developments will further improve missiles in the future. By the time you're an adult, it's likely there will be many new types of missiles, and they'll likely be even more effective at what they do. Even with the speed that new missile technology is being developed, it's important to know about the earliest missiles and how they came to be. By looking at these early examples, we can begin to understand how missiles became such an important—and destructive—weapon in warfare.

The First Ballistic Missile

The V-2 (Vergeltungswaffe Zwei, or "Vengeance Weapon Two") claims distinction as the world's first operational ballistic missile. A ballistic missile is one type of rocket that carries warheads to a specific target. Beginning in 1936, it was developed by a team of German scientists led by Wernher von Braun at Peenemünde, which is on an island in the North Sea. This missile was launched against Paris, France, on September 6, 1944, and then against Great Britain two days later on September 8, 1944.

The V-2 measured 47 feet (14 meters) long and weighed 28,000 to 29,000 pounds (12,700 to 13,150 kilograms). Its single-chamber rocket motor burned alcohol and liquid oxygen, and developed about 60,000 pounds (27,200 kg) of thrust. The V-2 carried a payload of 1,600 pounds (725 kg) of high explosives. It reached a height of almost 50 miles (80 kilometers) and a range of about 200 miles (320 km).

A. Bowdoin Van Riper, writing in *Rockets and Missiles*, tallied the damage inflicted by the V-1 (an early cruise missile) and V-2 missiles. Together, they destroyed 33,700 buildings and damaged 204,000, while killing 12,685 people and injuring 26,433.

V-1 and V-2 Rockets

German V-1 and V-2 rockets were launched against Great Britain during World War II (1939–1945). The V-1 ("V" for "Vergeltungswaffe" or "Vengeance Weapon") was called a "flying bomb." It was actually an early cruise missile. Its successor, the V-2, was a much more advanced missile. Together, they rained terror down upon Britons for more than nine months in 1944 and 1945.

In early 1945, as the Soviet army advanced toward Germany, von Braun and his team of scientists fled Peenemünde. They all went into hiding until they could surrender willingly to American troops. In a remarkable turn of events, von Braun signed a contract with the U.S. government. The Americans flew him to the United States, where he began a new chapter in his life.

Shown here is Peenemünde, Germany. This location played a huge part in the development of missile technology.

Cold War Missiles

At the end of World War II, the United States and the Soviet Union (also known as the USSR) entered into a missile development race. The intense rivalry of the two nations formed a part of what was to become known as the Cold War.

Both the United States and the Soviet Union imported German scientists who had helped to develop the V-2 missile. Wernher von Braun and his team were brought to the United States under Operation Paperclip, the secret intelligence program that brought Nazi scientists to the United States late in World War II and thereafter. According to Thomas Parrish in *The Cold War Encyclopedia*, the U.S. government contracted von Braun to "undertake such research, design, development, and other tasks associated with jet propulsion and guided missiles" as might be directed. The aim of both the United States and Soviet Union was to produce ballistic missiles for the delivery of **nuclear** warheads.

Developers of the first American and Soviet ballistic missiles used the V-2 as a starting point but steadily incorporated refinements. The Soviets produced a range of missiles designated R-1 through R-5. Introduced in 1956, the R-5 was the last of the line based on the V-2. It had a range of 700 miles (1,125 km) and carried a payload of 3,000 pounds (1,360 kg).

Meanwhile, von Braun and his team developed the Redstone missile. More powerful and more accurate than the V-2, it could deliver a 7,000-pound (3,175 kg) warhead on target within a radius of 1,000 feet (305 m). At the same time, the United States developed a compact, lightweight nuclear bomb. Combining the two, the Redstone became the world's first missile capable of delivering a nuclear bomb. Together, the Soviet R-5M and the Redstone formed the dividing line between the first and second generations of ballistic missiles.

Intercontinental Ballistic Missiles

The second generation of ballistic missiles began in August 1957. The Soviets announced the first successful launch of the R-7. Nicknamed the Semyorka (or "Seven"), it was the world's first intercontinental ballistic missile (ICBM). The R-7 was liquid-fueled, had a range of 3,000 miles (4,830 km), and carried a

The Redstone rocket was also known as "Old Reliable" because of its many diverse missions.

payload of 12,000 pounds (5,445 kg). It was the forerunner of the long-lasting Soyuz line of launch vehicles.

In October 1957, the Soviets shocked the world with the launch of *Sputnik 1,* the first artificial Earth-orbiting satellite. The Americans matched the Soviet achievement when the Redstone missile launched the *Explorer 1* satellite in January 1958.

To keep pace with the Soviet missile advances, the United States accelerated work on the Atlas—America's first ICBM—in March 1954 and began work on an alternate missile named Titan in May 1955. Both missiles were powered with liquid-fueled engines. The first successful Atlas launch occurred at Cape Canaveral, Florida, in December 1957. Titan I took to the air 14 months later in February 1959. Their operational lifespans soon gave way to the Titan II and Minuteman I in the early 1960s.

Minuteman I was the first American ICBM to use solid rather than liquid fuel. By 1965, the United States had retired all Atlas and Titan I missiles. Titan II and Minuteman I became the mainstay of America's strategic missile force. (Land-based heavy bombers—B-2 and B-52—and submarine-launched ballistic missiles [SLBMs] form the other two arms of the U.S. strategic nuclear triad, or group of three.) By 1974, all Minuteman I missiles had been replaced by either Minuteman II or Minuteman III missiles. The air force deactivated Titan II in 1982.

For a time, Americans enjoyed a superior nuclear force, but the Soviets soon achieved nuclear equality. A new race to develop an antiballistic missile (ABM) began. Toward the end of the administration of John F. Kennedy in the early 1960s, American strategists adopted a defense doctrine called **mutually assured destruction** (MAD). The thinking behind MAD was that no nuclear power would attack another knowing that it would mean certain death to both the attacked and the attacker.

Cuban Missile Crisis

Perhaps the most well-known military standoff involving missiles is the Cuban Missile Crisis. In October 1962, a high-flying

U.S. U-2 aircraft spotted a Soviet ballistic missile on a launch pad in western Cuba. President Kennedy placed a naval blockade around the island and confronted the Soviets. During this time, the world stood on the brink of a nuclear war.

After 13 terrifying days, Soviet premier Nikita Khrushchev—in a secret agreement revealed later—agreed to remove Soviet missiles from Cuba. In return, the United States agreed to remove its missiles from Turkey and not to invade Cuba. The confrontation led to the Nuclear Test Ban Treaty in August 1963.

Although the Cuban Missile Crisis lasted less than two weeks, it was a scary time for Americans—and the world as a whole. Cuba is only about 100 miles (161 km) from the southern tip of Florida. At such a close range, it posed a serious threat to the United States. While the Cold War was far from over, the Cuban Missile Crisis was a sort of turning point. In addition to the Nuclear Test Ban Treaty, a direct communication link between Washington and Moscow was put in place to help defuse similar situations.

Join the Debate

1. Why was the V-2 missile such a powerful weapon?
2. Why do you think it was so important to both the United States and Soviet Union to have the superior nuclear force?
3. Despite there being no actual attack, why do you think the Cuban Missile Crisis was so scary for the world?

The BGM-109G cruise missile, shown here, was developed by the U.S. Air Force toward the end of the Cold War.

CHAPTER 2

DEVELOPING CRUISE MISSILES

For as long as the Cold War was, many felt its end came quite suddenly. In the 1970s, the United States and the Soviet Union signed agreements to limit their nuclear weapons. When Mikhail Gorbachev took power in the Soviet Union in 1985, other big changes took place, most notably the Berlin Wall between East Germany and West Germany being torn down in 1989. Then, in 1991, the Soviet Union broke up and became 15 different countries. Many countries did away with Communist rule, and the Cold War ended. However, the nuclear arms race was still very much in existence.

When it comes to weapons of war, many countries still want to have the biggest, most destructive weapons. Even if they don't have plans to use them, they want to know they have the best ones, in case the time ever comes to use them. That means that even when the Cold War ended, the development of more powerful missiles certainly did not.

Missiles in Baghdad

In the early morning of March 20, 2003, a series of explosions shattered the stillness of Baghdad, Iraq. Huge clouds of smoke rose in the air, and the red glow of flames lit up the horizon. The blasts signaled the start of the "shock-and-awe" attack on the Iraqi capital by the United States and allied coalition forces at the outset of the Iraq War. It was spearheaded by around 320 Tomahawk cruise missiles launched from ships in the Red Sea and the Persian Gulf. The destruction they wrought was all but indescribable. Al Jazeera, an Arabic language news station, reported simply, "Baghdad is burning. What more can we say?"

A week later, in a windowless, sand-colored building in the country of Bahrain, navy planners and "targeteers" stared at computer screens and prepared to launch a daily attack of Tomahawks. The Fifth Fleet's Strike Center, where they worked, was surrounded by razor wire.

Some Tomahawk targets were decided by the Pentagon weeks or even months earlier as part of the Iraq War plan. Navy planners spent hours in preparation, studying those targets. They had to examine satellite photos, plot coordinates, select

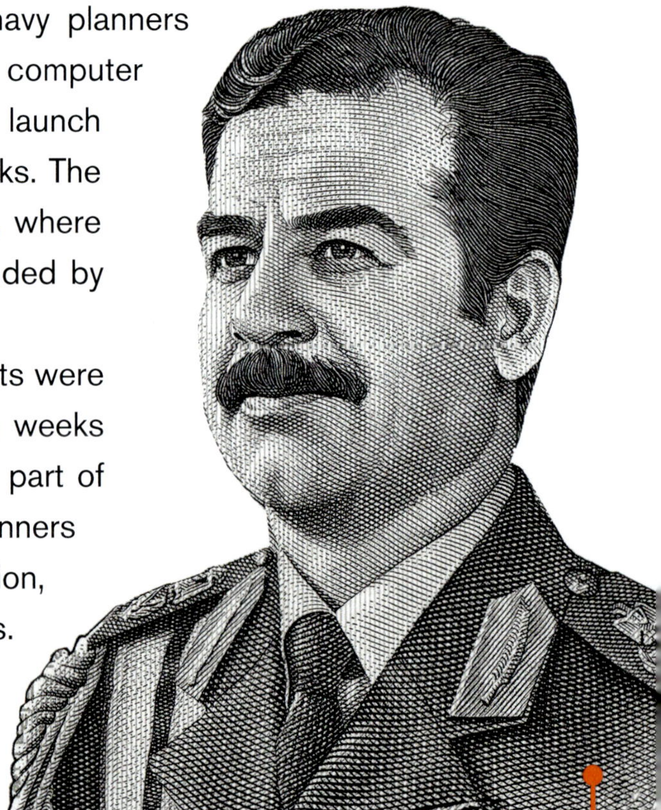

Saddam Hussein, president of Iraq from 1979 to 2003, was one of the targets of the U.S. missile attacks. However, he would not be captured until December 2003.

routes, and choose the ships or submarines that would launch the missiles.

In other cases, a target of opportunity demanding fast action—such as Iraqi president Saddam Hussein—suddenly presented itself and set off a chain reaction. After intelligence gathering and evaluation in Washington, D.C., President George W. Bush authorized a missile strike. His decision was sent electronically to the Strike Center in Bahrain. Officers there put together a target folder. Assessing the risk of civilian casualties, they decided how many missiles would be needed and which ships or submarines were best positioned to launch them. After alerting the air force to steer clear of missile flight paths, they were good to go.

When strike officers were satisfied that all their preparations were ready, they sent orders to ships operating in the Red Sea, Persian Gulf, or Mediterranean Sea. On each ship, the communications center rushed a copy of the order to its combat information center. There, planners fed the essential codes into the vessel's computer system, triggering an alarm. The sound of the alarm cleared the decks near the missile launch tubes. At the end of the chain, the ship's captain turned a key to activate the launch system, a launcher pressed a button, and a missile lifted off with a roar and began its deadly flight. In the first week of the Iraq War, the navy fired more than 800 cruise missiles.

Cruise Missile Development

Cruise missiles are so named because they generally operate at a cruising speed of 550 miles (885 km) per hour. The Tomahawk takes its name from the Native American war ax. Basically **derived** from the German V-1 and V-2 rockets of the World War II era, cruise missiles offer an alternative technology to ballistic missiles.

Cruise Missile Basics

Cruise missiles generally consist of a guidance system, a propulsion system, and a payload. They are housed in an airframe with small wings and a tail assembly for flight control. The airframe of the Tomahawk Block IV, for example, is divided into five increments: nose, payload, mid-body, aft-body, and tail-cone sections.

A cruise missile uses one or more guidance systems. These may include an inertial navigation system (INS), the **Global Positioning System** (GPS), terrain contour matching (TERCOM), or digital scene matching area correlation (DSMAC). INS uses stored data to follow a programmed route, while GPS transmits radio signals from Earth-orbiting satellites giving the location or distance from each satellite and the time of the transmission to pinpoint a location. TERCOM uses radar to detect and recognize landmarks as the missile flies over them. DSMAC operates like TERCOM but uses imagery instead of radar.

Cruise missiles are largely propelled by jet engines, usually using **turbofan engines** for their greater efficiency at low altitude and subsonic speed. For payloads, they use conventional or nuclear warheads.

By the end of World War II, the United States had 19 guided missile projects in progress. Many were based on the V-1 rocket. The 1970s gave rise to a new line of cruise missiles: the air-launched cruise missile, or ALCM (pronounced al-cum); the

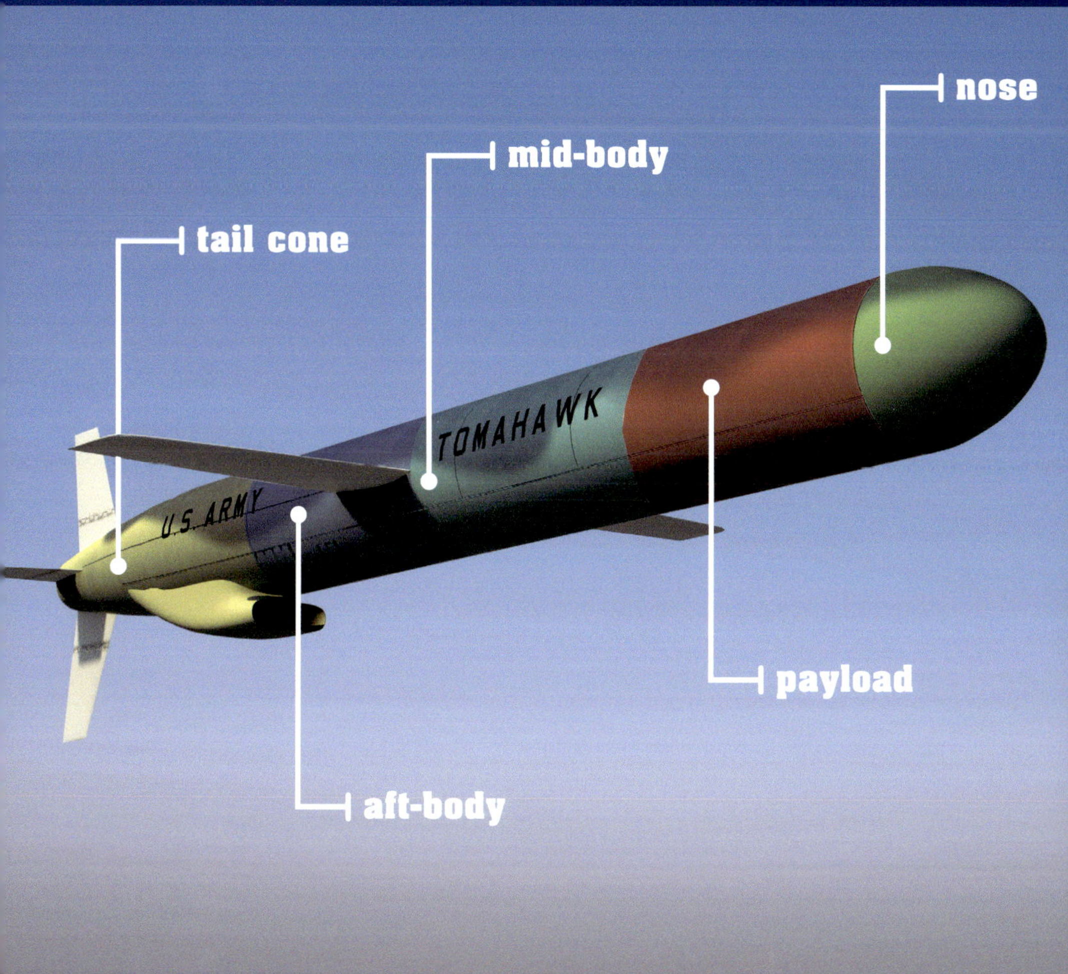

This diagram shows the five increments of a Tomahawk Block IV.

ground-launched cruise missile, or GLCM (glick-em); and the sea-launched cruise missile, or SLCM (slick-em). The ever-reliable Tomahawk evolved from the SLCM. It later morphed into today's Tomahawk Land Attack Missile, or TLAM. This is an all-weather, long-range, subsonic cruise missile. It is used for land-attack warfare and is launched from U.S. Navy surface ships and U.S. Navy and Royal Navy (United Kingdom) submarines.

The TLAM is designed to hit stationary or moving targets on land.

The Stealthy Missile

Cruise missiles are known for their stealth. They are able to travel closer to the ground than other types of missiles and therefore can stay out of view of most radar systems. Their power, accuracy, and ability to fly relatively undetected make cruise missiles a valuable weapon to have in any fleet.

Join the Debate

1. Do you think cruise or ballistic missiles are more powerful? Why?
2. Should there be a limit on how many missiles a military can have? Why or why not?
3. How does one military having missiles affect other militaries around the world?

The AGM-114 Hellfire is a laser-guided missile known for being able to defeat any enemy tank.

CHAPTER 3

ALL ABOUT CONVENTIONAL GUIDED MISSILES

The United States' arsenal of conventional guided missiles covers a broad spectrum of applications, both offensive and defensive. Missiles fall into three basic categories. Variations of each type increase the total kinds of missiles to more than 20.

In recent years, conventional guided missiles have played key roles in important military operations overseas, particularly during the Gulf War (1990–1991).

Operation Desert Storm

On the night of January 16, 1991, four massive MH-53J Pave Low helicopters lifted off a chilly desert airfield in Saudi Arabia into a moonless night. Pilots of the 20th Special Operations Squadron could see their breath forming in foggy puffs in the cockpits of the MH-53J choppers. Right behind them, eight AH-64 Apache attack helicopters and a Blackhawk helicopter followed them into the pitch-black sky. Collectively, they formed Task Force Normandy, commanded by Lieutenant Colonel Dick Cody. Their mission was to take out the Iraqi early warning network.

The AH-64 Apache attack helicopter, shown here, is one type of U.S. Army aircraft that commonly carries AGM-114 Hellfire missiles.

The Apaches and the Blackhawk belonged to the 101st Airborne Division. Troops of the Special Ops group were along as extra insurance in the event the choppers crashed or were shot down. The Blackhawk carried mechanics and spare parts.

For the next two hours, they all skimmed along over the desert at sand-dune height toward the Iraqi early warning net and two designated targets. At preset coordinates, the group peeled off into two attack teams. The teams were ordered to maintain absolute radio silence until 10 seconds before opening fire.

Early in the morning of January 17, 1991, the eight Apaches zeroed in on their targets and hovered low and unseen, 4 miles (6.4 km) south of the radars. Images of two Iraqi radar sites appeared on their forward-looking infrared screens. The sites were positioned just north of Saudi Arabia to detect intruding aircraft. They maintained constant communication with four Iraqi fighter bases, as well as a direct line to the Intelligence Operations Center in Baghdad.

First Lieutenant Tom Drew, at the controls of Number 976 chopper, broke radio silence. “Party in ten,” he said, as recounted by Richard Mackenzie in *Air Force Magazine*. Ten seconds later, on a prearranged cue, the Apaches fired off a volley of 27 AGM-114 Hellfire air-to-ground missiles. Their shots announced the start of Operation Desert Storm and touched off U.S. involvement in the Gulf War. Task Force Normandy destroyed 16 separate radar installations at each of two sites. Nothing but piles of twisted metal remained at the sites—and a 20-mile-wide (32 km) pathway was now open all the way to Baghdad.

Twenty-two minutes after the Apaches completed their lethal work, a fleet of about 100 allied aircraft arrived at Iraq’s devastated portal and soared through the huge gap in the Iraqi surveillance network. Racing north, they dispensed similar destruction on critical, first-night targets in Baghdad. They found the city already under attack by U.S. Air Force F-117 stealth fighter-bombers. The F-117s had used their stealth advantage to slip through Iraqi airspace without detection. They started their attack soon after the Apaches fired their opening rounds.

America's Missile Arsenal

One of the most widely used offensive missiles in the U.S. inventory is the AGM-114 Hellfire. It is an air-to-surface missile originally designed as an anti-tank weapon. The Hellfire has a length of 5 feet, 4 inches (162.56 centimeters) and a diameter of 7 inches (17.78 cm), and it carries a 16-pound (7.26 kg) high-explosive, shaped charge. Though generally fired from Apache or Super Cobra attack helicopters and Predator unmanned combat air vehicles, it can also be launched from fixed-wing aircraft, ships, and land-based systems.

In the Gulf War, U.S. and Israeli Patriot weapons attempted to intercept 44 extended-range SCUD missiles launched by Iraq against Israel and Saudi Arabia. American claims of success were eventually revised downward from a hit rate of 96 percent to 61 percent. Its record was still efficient enough for the Patriot to earn the nickname "Scudbuster." The latest version of the Patriot promises a far greater success rate.

Another defensive weapon is the Standard SM-2ER Block IV missile (ER means "extended range"). It is an all-weather, supersonic, ship-launched, medium-to-long-range fleet air defense missile. Also known as the RIM-156A, it provides defense for

Patriot Missiles Today

The countries of Ukraine and Russia share a border. On February 24, 2022, Russia invaded Ukraine. Millions of Ukrainians fled their country, and many were displaced. In March, the United States sent two Patriot missile systems to Poland in an effort to deter Russia. As of the **deployment**, Ukraine was not a member of the **North Atlantic Treaty Organization** (NATO), but Poland was. The U.S. government stated that it deployed the missiles for defense purposes in case the conflict were to carry over to Poland or other nearby NATO nations.

At the time, Russia had deployed missiles of its own to areas bordering Poland. Retired U.S. Army general Wesley Clark, a former NATO commander, said that the Patriot system could intercept Russian missiles if necessary.

In 2003, a Patriot missile system was deployed in Operation Iraqi Freedom. It succeeded in hitting nine incoming missiles. However, it was also responsible for several incidents of friendly fire. Friendly fire means a weapon from one's own side is fired at its own troops, usually causing accidental death or injury to its own forces.

Patriot is an acronym. It stands for Phased Array Tracking Radar to Intercept of Target.

an entire fleet area. (The R means it's surface-ship launched; IM stands for "intercept missile.") It is deployed on destroyers and cruisers and is the latest version of the RIM-66/67 series to enter the fleet as of 2022.

Sailors from the U.S. Navy are shown here working on a RIM-7 Sea Sparrow missile.

Guiding a Missile to Its Target

A guided missile is being controlled the entire time it is moving toward its target. Today's guided missiles use the latest technology for the most accurate control. Some missiles can even be guided to hit moving targets. As the target's path moves, so too does the missile's path. While many missiles today are guided by computer systems, the people working with the missiles need to be extremely careful and precise in everything they do.

Join the Debate

1. Would you ever want to fly a helicopter carrying a missile? Why or why not?
2. Do you think the risk of friendly fire is worth the protection Patriot missiles may offer?
3. Do you think it was right for the United States to deploy Patriot missiles to Poland in 2022?

Smart bombs have fins or wings that move, allowing for adjustments to be made to the way the bomb is falling.

CHAPTER 4

MAKING SMARTER WEAPONS

As technology has advanced, everything has gotten "smarter." We have smartphones, smart TVs, smart refrigerators, and even entire smart homes. Of course, the military has smart technology, too, namely smart bombs, also called smart missiles. In terms of technology, "smart" usually means the most high-tech or advanced, usually involving some kind of automation. The U.S. military has some of the best minds and top STEM experts working on creating the best smart weapons.

Using Smart Weapons in Kobani

Early in America's air campaign against the Islamic State in Iraq and Syria (ISIS) that began in August 2014, B-1 crews from the 9th Bomb Squadron at Dyess Air Force Base, Texas, focused on a single town in Syria called Kobani. Kurdish forces known as Peshmerga were trapped in their own city under the advance of the **insurgents**.

ISIS fighters moving openly on top of buildings and bridges provided plenty of targets for the airmen. The B-1 Lancer was made for missions just like this one. It carried a wide variety of weaponry, including both general-purpose and smart bombs.

Few targets can survive its destructive force. Additionally, the B-1 can stay over the target area for about 10 hours without needing to refuel. This loitering capability allows U.S. airmen to react to targets supplied by Kurdish forces or develop their own targets to drive off the insurgents.

The squadron dropped an estimated 660 bombs on Kobani, killing more than a thousand ISIS fighters. It delivered about one-third of all the bombs dropped during the first five months of Operation Inherent Resolve, the official name of the air campaign against ISIS.

The 9th Bomb Squadron completed its deployment in late January 2015. By then, the Kurdish Peshmerga had secured Kobani and claimed victory. "To be part of something, to go out and stomp these guys out," said a B-1 pilot named Major Johnson, "it was completely overwhelming and exciting."

Sending Smart Bombs

President Barack Obama authorized the use of "smart bombs" in Iraq and Syria in the summer of 2014. By mid-March 2015, U.S. and coalition warplanes had dropped more than 10,200 precision-guided bombs on ISIS forces. Almost every bomb used by the United States in the air war against ISIS was guided by a satellite or laser. By comparison, only 10 percent of the bombs used in the Gulf War in 1991 had advanced systems.

Advanced smart bombs are constructed from conventional—or "dumb"—bombs by attaching kits to them. Various conventional bombs are used to convert dumb bombs to smart bombs, including the Mk 82–84 series, the BLU-109 Penetrator, and the BLU-116 Advanced Unitary Penetrator.

President Obama officially sanctioned the use of smart missiles in Iraq and Syria in 2014.

More About Smart Bombs

"Precision-guided munitions" is the government phrase for smart weapons. A precision-guided munition (PGM) is a missile, bomb, or artillery shell that is fitted with a terminal guidance system. The guidance system consists of electrical equipment that guides it during its last phase before impact. In short, a smart weapon is a projectile with a homing device. PGMs are designed to precisely hit a specified target while minimizing collateral damage (unintentional death, injury, or property damage).

Conventional bombs simply consist of an explosive material packed in a sturdy case with a fuse mechanism. The fuse contains a triggering device that ignites the explosive, causing the bomb to go off. Conventional bombs are considered "dumb" because they fall freely to the ground without any way to steer them. Consequently, many dumb bombs might be required to take out a single target. By contrast, smart bombs can be controlled in flight to hit the smallest target with pinpoint accuracy.

To keep pace with the daily demand for smart munitions, the U.S. Air Force formed a special "ammo team" at Dyess Air Force Base in Texas. The crew upgraded 500- and 2,000-pound (225 and 910 kg) free-falling bombs to smart-bomb status by affixing a dozen parts to each "dumb" bomb—fuses, radar-equipped nose cones, GPS-guided tail fins, and more.

Crews at Dyess Air Force Base, shown here, have upgraded many "dumb" bombs to smart bombs.

Guiding Smart Bombs

Smart bombs currently used by U.S. forces use several guidance systems. These systems are electro-optical, laser, or infrared. With an electro-optical system, a camera in the nose of the bomb sends images of the target to the crew. The crew can then guide the bomb to the target. With laser-guided bombs, a laser beam is pointed at the target. The bomb has sensors that follow the beam to the target. Last, smart bombs with infrared guidance systems react to heat from the target to guide the bomb toward impact.

Most commonly, a smart bomb's guidance system is attached to the nose of the missile.

Join the Debate

1. Why are smart bombs superior to "dumb" bombs?
2. What are some drawbacks to smart bombs?
3. Which of the three guidance systems for smart bombs seems most accurate? Why do you think so?

Air force crews responsible for missile operations often work behind the scenes making very important decisions.

CHAPTER 5

WORKING WITH MISSILES

There is an array of jobs in the U.S. military that deal with missiles. In the U.S. Air Force, nuclear and missile operations officers are responsible for running and managing the country's nuclear and missile capabilities. These officers are entrusted with the last line of defense of the United States. Ultimately, their decisions can affect the fate of the world.

The Missileers

Since the 1960s, a small group of air force personnel known as missileers has been working in a shadowy world deep underground at U.S. launch control centers, responsible for 450 Minuteman III missiles. Missileers remain vigilant 24/7, ready to act at once on the president's order—an order most hope will never come. Their unique job allows no room for error. It demands efficiency, focus, and constant operational alertness. To meet the uncommon physical and mental demands of their responsibilities, missileers must undergo rigorous initial and recurrent training.

Training Time

Future managers of America's Minuteman III ICBM system receive operational readiness training at Vandenberg Space Force Base. The base is located on the central coast of California between the towns of Lompoc and Santa Maria. Training is conducted by the 532nd Training Squadron of the 381st Training Group.

The LGM-30 Minuteman III (*shown here*) is the longest-serving ICMB in the United States' weapons arsenal. It replaced the earlier Atlas and Titan. The LGM-30 was itself replaced in part by the LGM-118 Peacekeeper, or MX, missile, but then replaced its replacement, primarily due to cost considerations.

Initial and advanced training is provided in three phases: initial qualification training, unit qualification training, and recurrent training. Trainees are taught ICBM operations and procedures and are trained as electronic, electro- and missile-mechanical, and facility maintenance technicians.

After six months of initial qualification training, the new missileers are assigned to missile combat crews at one of three Minuteman III sites, where they continue unit training. Thereafter, crews undergo training and evaluation once or twice a month to ensure their skills remain razor sharp. Monthly training consists of about two days of classroom training, covering such things as emergency war orders, weapons systems operations, and code controls. Periodic competitions are also held with other crews to instill pride in their unit and vie for top honors.

A Great Responsibility

In addition to going through rigorous training to get to the position, a missileer needs to be willing to bear a great responsibility. One toggle switch can send a single nuclear warhead—with 20 times the explosive force of the first atom bomb—hurtling across the globe at speeds of 15,000 miles (24,140 km) per hour. Luckily, in more than 50 years of duty, Minuteman missileers have never had to launch a live missile, but they are ready at any moment. In fact, they showcased their readiness in August 2021. Airmen from Minot Air Force Base in North Dakota, as well as Malmstrom Air Force Base and F. E. Warren Air Force Base, launched an unarmed Minuteman III missile in California. The test launch showed that although we hope to never need it, the air force is ready to launch such a weapon at any moment.

A Day in the Life

A typical day for two launch control officers of the oncoming watch at Minot Air Force Base begins on an elevator that takes them dozens of feet underground. They exit at the bottom of the elevator shaft and proceed through a sealed 8-ton (7.25 metric tons), 2-foot-thick (0.61 m) concrete blast door, arriving at the Launch Control Center (LCC). The LCC is called the "capsule" because its shape resembles that of a pill. It is suspended in air by four king-sized shock absorbers. A second blast door seals its entrance. The new watch officers announce their arrival, and the door swings open. They are greeted by the old watch officers and enter the capsule. Crammed with green and white computer equipment, the capsule's tight interior speaks to the high-tech nature of the nation's work being conducted there.

LCC procedures require old and new launch commanders to conduct detailed inspections of the entire capsule and all its equipment. They verify that everything is in order and all the equipment is working properly. After a satisfactory inspection, the incoming commander signs a form attesting to this satisfactory inspection and takes command of the LCC. For the next 24 hours, this person and their deputy will remain in charge of the LCC and 10 Minuteman III ICBMs that are dispersed about the North Dakota countryside in strategic locations. Above ground, a support team stands by to provide security and food and to ensure that all equipment remains in working order.

The "capsule" is located underground at Minot Air Force Base.

TIMELINE OF MILITARY MISSILES

1944: Unmanned V-1 and V-2 rockets bomb Great Britain during World War II.

1957: Soviets shock the world with the launch of *Sputnik 1*; Americans launch the Atlas intercontinental ballistic missile.

1958: Americans launch the *Explorer 1* satellite with a Redstone missile.

1959: The Titan I missile takes to the air.

1962: U.S. U-2 aircraft spots a Soviet ballistic missile in western Cuba, inciting the Cuban Missile Crisis.

1965: Atlas and Titan I missiles are replaced by Titan II and Minuteman I missiles.

1974: Minuteman I missiles are replaced by the Minuteman II and Minuteman III missiles.

1982: U.S. Air Force deactivates Titan II missiles.

2003: U.S. cruise missiles strike Baghdad at start of the Iraq War.

2014: President Obama orders air strikes against ISIS in August.

2022: The United States deploys two Patriot missile systems to Poland after the Russian invasion of Ukraine.

This timeline shows how missiles have been used by militaries in the 20th century and beyond.

Join the Debate

1. Would you ever want to be a missileer?
2. What do you think the hardest part of being a missileer would be?
3. Missiles are extremely dangerous weapons. Do you think it's important for the military to have them ready to use at all times? Why or why not?

GLOSSARY

ballistic missile: A missile (rocket) that follows a high-arc trajectory to deliver one or more warheads to a designated target.

deployment: The act of organizing and sending people or tools for a certain purpose.

derive: To come from something else.

Global Positioning System: A radio navigation system that uses multiple satellites to allow land, sea, and airborne users to determine their exact location, velocity, and time anywhere on the earth.

insurgent: One who rises in revolt against civil authority or an established government; a rebel.

mutually assured destruction: A defense policy assuring no nuclear power would attack another, for it would mean certain death for both.

North Atlantic Treaty Organization: An organization of countries that have agreed to provide each other with military support.

nuclear: Having to do with energy created when the nuclei of atoms are split apart or joined together.

propel: To push or direct something in a certain direction.

turbofan engine: A type of aircraft jet engine based on a gas turbine that generates thrust by using a combination of a ducted fan and a jet exhaust nozzle. It passes part of the air from the ducted fan through the core, which provides oxygen to burn fuel to create power.

FIND OUT MORE

Books

Duncan, Terri Kaye. *Thirteen Days of Tension: The Cuban Missile Crisis*. New York, NY: Rosen Publishing, 2021.

Hamilton, John. *Missiles and Spy Satellites*. Minneapolis, MN: Abdo Publishing, 2019.

Ripley, Tim. *Torpedoes, Missiles, and Cannons: Physics Goes to War*. Minneapolis, MN: Lerner Publications, 2018.

Websites

Cuban Missile Crisis
history.state.gov/milestones/1961-1968/cuban-missile-crisis
Read more about the Cuban Missile Crisis on the website of the United States Office of the Historian.

LGM-30G Minuteman III
www.af.mil/About-Us/Fact-Sheets/Display/Article/104466/lgm-30g-minuteman-iii/
Learn more about this powerful missile on the U.S. Air Force website.

Organizations

Smithsonian National Air and Space Museum
655 Jefferson Drive SW
Washington, DC 20560
airandspace.si.edu
The Air and Space Museum has information and exhibits on all types of military flight. Its collections include missiles and aircraft that have carried missiles.

U.S. Air Force
1670 Air Force Pentagon
Washington, DC 20330-1670
www.airforce.com
Browse the U.S. Air Force's website to find out more about this branch of the military, including information about the missiles it deploys and the people who work on them.

Publisher's note to educators and parents: Our editors have carefully reviewed these websites to ensure that they are suitable for students. Many websites change frequently, however, and we cannot guarantee that a site's future contents will continue to meet our high standards of quality and educational value. Be advised that students should be closely supervised whenever they access the internet.

INDEX

ABOUT THE AUTHOR

Tanner Billings is a writer living in New York. He enjoys studying the history of the military and keeping up with modern military technologies. When he is not writing, he enjoys hiking and taking trips to national parks with his family and dogs.